12/04

$26.50

D0572787

THE LIBRARY OF SATELLITES

EARTH IMAGING SATELLITES

JAN GOLDBERG

the rosen publishing group's
rosen
central

Published in 2003 by The Rosen Publishing Group, Inc.
29 East 21st Street, New York, NY 10010

First Edition

Library of Congress Cataloging-in-Publication Data

Goldberg, Jan.
Earth imaging satellites / Jan Goldberg.— 1st ed.
   p. cm. — (The library of satellites)
Summary: Examines the technology involved in earth imaging satellites, the services they provide, the tasks they can perform, the history of these satellites, and their likely future applications.
Includes bibliographical references and index.
ISBN 0-8239-3853-0
1. Scientific satellites—Juvenile literature. 2. Remote sensing—Juvenile literature. [1. Scientific satellites. 2. Artificial satellites.]
I. Title. II. Series.
TL798.S3 G65 2003
550'.28'4—dc21

2002012856

Manufactured in the United States of America

CONTENTS

INTRODUCTION

The huge snow tractor rumbled slowly across the frozen surface of Antarctica. Traveling inland from McMurdo Station (a United States Antarctic research center), it headed toward the center of this icy land. The tractor was dragging a large shed filled with sixteen tons of explosives. Supplies and equipment are usually flown into and out of McMurdo Station on transport planes. This cargo, however, was much too large and far too dangerous to deliver by plane.

Suddenly, without warning, the tractor hit a hidden crevasse, or deep crack, in the thick ice. The tractor—and the two men driving it—fell down into the crevasse. One of the tractor's windows shattered into pieces when it hit the bottom. Glass, snow, and ice poured into the cab section of the tractor, where the drivers were sitting. The huge shed full of explosives balanced precariously on the icy ledge high above them. The two men waited there for several hours while the tractor's passengers tried to get them out. Finally, the men were pulled to safety with ropes.

The icy continent of Antarctica is covered with many crevasses like the one that swallowed the snow tractor. Unfortunately, these dangerous cracks in the ice cannot

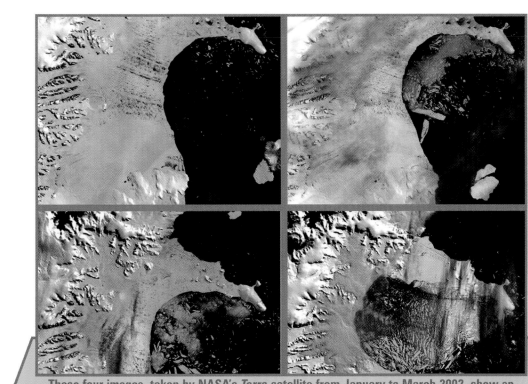

These four images, taken by NASA's *Terra* satellite from January to March 2002, show an ice shelf shattering into thousands of icebergs as it breaks away from Antarctica. Once detached, the remains of the Rhode Island–sized shelf dissolved quickly, causing the sea level to rise.

be seen by someone on the ground. They are usually hidden from view by large snowdrifts. Crews flying in airplanes close to the ground cannot see them, either. The only set of eyes that can detect these cracks in the ice are not human. Instead, they belong to artificial satellites that orbit Earth.

Since 1991, when this accident happened, scientists have been working to find safe traveling routes across Antarctica, a continent that is in an almost constant state of flux. Because of its surface of snow and ice and its harsh weather, Antarctica's topography is always changing. Dr. Carolyn Merry, a satellite imagery and surveying expert, and her team of researchers have studied satellite images of the entire continent in an attempt to map out many safe transport routes that do not go near any dangerous hidden cracks. Supplies and equipment that cannot be loaded onto a plane now have a way to get from one place to another on Antarctica. The detailed satellite images that the scientists studied have probably saved many lives.

As you read this, there are thousands of satellites orbiting Earth. There are many types of satellites, including meteorological satellites (that help predict weather), communications satellites (that enable radio and television broadcasts, cell phone calls, Internet access, and e-mail messages), and military satellites (that can act as high-altitude spies).

This book is about a type of satellite called an Earth imaging satellite. Dr. Merry and her team used an Earth imaging satellite to map the entire surface of Antarctica. Earth imaging satellites are offering us previously unseen glimpses of our world and a far greater understanding of the planet, its inhabitants, and its atmosphere. More important, they are alerting us to the environmental dangers we face—from floods and hurricanes to pollution and global warming—and offering us insights on how to confront these problems.

CHAPTER ONE

SATELLITE HISTORY AND OPERATIONS

A satellite is any object in space that is orbiting, or traveling around, another object in space. For example, Earth is a satellite of the Sun. The Moon is a satellite of Earth. Earth and the Moon are called natural satellites. When we first learned how to put man-made objects into orbit around Earth, these objects were called artificial satellites. Over the years, artificial satellites have become more and more common. Also, their equipment has become much more advanced. Artificial satellites eventually came to be referred to simply as satellites.

SPUTNIK AND THE RACE FOR SPACE

The former Soviet Union (a Communist empire that broke apart in 1991 into several independent nations, including Russia) sent Earth's first artificial satellite into orbit. The 184-pound satellite was named *Sputnik I*, which means "traveling companion." *Sputnik I* was launched on October 4, 1957. *Sputnik II*, which was launched a month later,

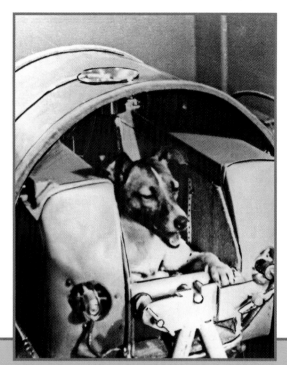

Unfortunately, the competitive political pressure of the Cold War forced *Sputnik II* to be built in such a hurry that the engineers did not have time to develop a life-support system adequate enough for Laika, the first living being in space, to survive her trip.

carried the first living creature into space—a dog named Laika.

After seeing what the Soviet satellites could do, the United States worked quickly to finish building a satellite of its own. At this time, the United States and the Soviet Union were locked in a tense standoff known as the Cold War, so-called because the simmering hostility never led to actual warfare. The two sides never exchanged gunfire or missiles, but they each supported and funded combatants in other warring nations. By supporting these "proxy" wars, each superpower hoped to spread its influence and ideology—American capitalism and democracy vs. Soviet Communism—around the world.

The race to reach space was an outgrowth of this Cold War strategizing. The first country to reach space and maintain a technological edge there would not only score a huge public relations victory, but would also gain a new "high ground" from which to spy on, defend itself against, or even attack its enemies. The Sputnik launches represented a humiliating failure of American preparedness and enterprise, as the nation watched the launch with the rest of the world in awe and apprehension. Never again did the United States want to be caught unaware, a mere spectator to great events. It vowed to not only catch up with the Soviets, but also to win the space race.

On January 31, 1958, the United States launched its first artificial satellite, called *Explorer I*. It had finally established its presence in space and would quickly begin to lead the way in manned flights, space research and exploration, and satellite technology. Within just a few years of the Sputnik and Explorer series launches, many different satellites—communications, reconnaissance (spy), weather, Earth imaging—were in orbit high above Earth.

ROCKETS

Sputnik I was a relatively crude satellite. It was little more than a metal ball that made an electronic beeping noise as it circled Earth, at the mercy of gravity and centrifugal force (both of which, working together, keep a satellite spinning around Earth at a certain altitude and orbit). The real triumph of the Sputnik program was in figuring out a way to get the satellite into orbit in the first place. The development

"The Earth is the cradle of the mind, but we cannot live forever in a cradle," wrote Konstantin Tsiolkovsky in a letter to a friend. In Kaluga, Russia, a statue stands to honor his great achievement as the designer of the first rocket.

of a rocket powerful enough to lift an object into orbit was a long time in the making.

The history of rockets actually goes all the way back to China, in the 1200s. The Chinese learned how to pack gunpowder into a cardboard cylinder. Then they would light a fuse attached to the end of the cylinder. When the gunpowder ignited, gases formed. These gases then pushed backward in the cylinder, causing the cylinder to move forward.

Around 1903, a Russian scientist named Konstantin Eduardovich Tsiolkovsky (1857–1935) first suggested the idea of space travel. He thought that if we could send man-made objects into outer space, those objects might be able to tell us what it was like up there. He suggested that rockets were one possible way for objects to be sent into space. He also thought that liquid oxygen and

THE FATHER OF AMERICAN ROCKETRY

An American scientist named Robert H. Goddard carried on the work of Tsiolkovsky. In 1926, Dr. Goddard launched the first rocket that contained liquid gases instead of gunpowder while working in his backyard in Massachusetts. His rocket went only 40 feet (12.2 meters) into the air, spending about two and a half seconds in flight. But Dr. Goddard's important experiments led directly to the advanced rockets that are used today. In addition, he developed an automatic steering system for his rockets. As a result, Dr. Goddard is often referred to as the Father of American Rocketry. The National Aeronautics and Space Administration (NASA), a U.S. government agency devoted to space exploration and technology, named its Goddard Space Flight Center (GSFC) in his honor. The GSFC is located in Greenbelt, Maryland, just outside of Washington, D.C. It is an organization of scientists and engineers dedicated to research of Earth, the solar system, and the universe.

hydrogen could create a more powerful and effective boost for a rocket than gunpowder.

LAUNCHING A SATELLITE

Many satellites are launched into space with rockets. Some are taken into space in the cargo bay of space shuttles, from which they are carefully removed and sent into space by shuttle astronauts. Even the space shuttles need powerful rockets to get into space, however. Rockets are the only things we have invented so far that are strong and fast enough to escape Earth's gravity. Gravity is the strong force that keeps us rooted to Earth. Without it, we would not be able to remain standing on a planet that is spinning at about a thousand miles (1,609 km) an hour. Gravity keeps the Moon revolving around Earth and Earth around the Sun in orderly and predictable orbits. Without gravity, the planets and stars would spin off and collide in an explosion of random chaos. Gravity pulls objects down, toward the center of the planet (which is why when we jump, we always come back down to Earth again). So in order for an object to escape the heavy force of gravity of Earth's atmosphere, it has to have enough speed and power to push through to a high altitude where there is far less gravity.

A rocket that carries a satellite into space is known as a multistage rocket. This means that there are two or three stages, or separate units, that make up each rocket. When the fuel from one stage is used up, that stage separates from the rest of the rocket. Then, the next stage fires up and takes over. The used-up section of the rocket normally falls back to Earth. If all goes as planned, it lands far out in the middle of an ocean. As the different stages fall away and the rocket becomes lighter, not as much fuel is needed to power it.

Once the launch vehicle is in the thinner air of outer space, about 120 miles (193 km) up, a guidance system within the main rocket is used to tilt it eastward. Earth rotates to the east, and a rocket heading in the same direction gets an extra push and a little more speed from the force of Earth's rotation. When it finally reaches the altitude and speed at which

At California's Vandeberg AFB, a Boeing Delta II rocket clears the launch pad with the *Landsat-7* satellite as cargo. Only a few hours and hundreds of miles later, *Landsat-7* was producing detailed photographs of Earth.

the satellite was designed to work, the rocket releases its cargo. The satellite then flies free of the spent rocket.

IN ORBIT

Having arrived at the right altitude, the satellite is then placed into orbit (an orbit is one object's path around another object). Though it has to first escape gravity to get into space, once in space a satellite then depends on gravity to keep it in orbit.

A satellite can revolve around the globe because of two factors: inertia and gravity. Inertia is the tendency of any physical object to either stay still or to keep moving in a certain direction and at a certain speed until an outside force interferes with the object's speed and direction. A satellite, whether natural or artificial, is moving all the time as it spins alongside Earth. There is virtually no object in space to make it stop moving or alter its direction. What keeps it from flying way beyond Earth into deepest space is gravity. Gravity pulls it toward Earth at the same time that inertia propels it away from Earth. When these two factors are balanced, the satellite keeps moving but stays "tied" to Earth; it orbits Earth in a regular and predictable circular motion.

For a satellite to achieve the proper balance between inertia and gravity, the satellite has to move at just the right speed. If it moves too fast, the force of motion will propel it beyond the power of gravity to hold it in a stable orbit. If the satellite moves too slowly, gravity pulls it down out of orbit, and it crashes into Earth or burns up in Earth's atmosphere. The correct speed—the speed that allows it to stay securely in orbit—is called orbital velocity. The exact speed necessary to keep a satellite orbiting the planet depends on how far the satellite is from Earth and what type of orbit it maintains.

Satellites following a polar orbit are about 540 miles (869 km) above Earth. A polar orbit follows a path around Earth from the North Pole to the South Pole, then around the other side. Satellites following this type of orbit circle Earth about eighteen times per day, allowing them to glimpse the entire surface of Earth with their "eyes" several times a day. Polar orbiting satellites get a closer view of Earth than satellites that are orbiting higher up. They can monitor remote locations, such as the Arctic and the Antarctic. Earth imaging satellites and many weather satellites follow a polar orbit. In fact, most follow a special type of polar orbit called a sun-synchronous, or sun-synchronized, polar orbit. This means that the satellite crosses over the equator at the same local time every day.

Another type of orbit is called a geostationary, or geosynchronous, orbit. A satellite following this type of orbit moves directly above the equator. Since it orbits at the same speed as Earth's rotation, it stays over the same location at all times. A satellite in a geostationary orbit must be 22,310 miles (35,905 km) above Earth to stay in orbit. At this altitude, the satellite's orbital velocity, or speed, will match Earth's, allowing the satellite to remain hovering over a fixed point. Most communications satellites and certain weather satellites follow this type of orbit.

In a 1945 article for *Wireless World,* science fiction author Arthur C. Clarke hypothesized that geosynchronous satellites would exist within the next fifty years. In 1960, the Hughes Corporation urged NASA to attempt to design a better communications satellite. Twenty-two months later, *Syncom II*, the world's first geosynchronous satellite, was in orbit over Earth.

SOLAR ENERGY

Though there are many different types of artificial satellites, they all have some important things in common. For example, they are all powered the same way. Out in space, there is no electricity like we use here on Earth to power our TVs, refrigerators, and lamps. The energy of the Sun is readily available, however, and satellites use this energy to power their instruments. Most satellites are covered with solar cells. When the Sun shines on a satellite, the solar cells convert the Sun's energy into electricity. This type of energy is called solar energy. The cells also give power to the storage batteries in the satellite. These batteries store solar energy for the periods when Earth comes between the satellite and the Sun, throwing the satellite into darkness. Without the storage batteries, the satellite's operations would shut down with the onset of darkness.

SENSORS

All satellites have sensors, which are devices that can be programmed to detect and measure various physical properties of Earth below. These sensors can detect the presence of and identify water, gases, minerals, and chemicals. The information gathered by sensors can be used to help mining companies detect the presence of coal or meteorologists predict the likelihood of floods or estimate snowfall accumulations. Satellite sensors can take digital pictures of the terrain or oceans on Earth. The sensors can then transmit the data and images they have collected down to ground stations on Earth using their radios, antennas, and onboard computers. This information can then be studied and interpreted by computers and scientists. The antennas also allow the sensors to receive instructions and messages sent to them by the ground station.

AN EARLY OBSERVATION SATELLITE

On April 1, 1960, NASA launched its first meteorological, or weather, satellite into a sun-synchronous polar orbit. They named it *TIROS*, which stands for Television Infrared Observation Satellite. *TIROS* observed cloud formations, movements, and circulation patterns, making predictions of weather patterns easier. Improvements to the TIROS satellites led to a series of weather satellites named Nimbus.

The Nimbus satellites were much more complex than the TIROS satellites. They featured an improved television camera system and infrared sensors, which allowed for nighttime observations. The TIROS satellites' instruments spun with the satellites so that when the satellites were not facing Earth, the instruments could not make any observations. The Nimbus instruments, however, were mounted on a rotating axis, so that when the satellites faced away from Earth, their instruments could still face the planet and continue making observations. In this way, the Nimbus satellites were able to monitor weather patterns continuously.

By carefully studying weather images sent back from TIROS and Nimbus satellites, scientists soon realized something very important. Studying just a few days' worth of satellite images helped them greatly in predicting weather patterns around the world. But if they pieced these images together so that they had months and years worth of images, they could monitor such important things as the patterns of the ocean currents, changes in ocean

The Television Infrared Observation Satellite (*TIROS*) became the United States's first weather satellite. Weighing a hefty 270 pounds (123 kilograms), *TIROS* tested experimental television techniques for monitoring weather.

temperature, the melting of sea ice, the erosion of the coastlines, and even the growth of cities.

NASA's next satellite project would build on this discovery and begin to construct an awe-inspiring "photo album" of Earth's surface as it changed from year to year.

CHAPTER TWO

LANDSAT

With the effectiveness of satellite observations of weather well established, NASA got to work on a new family of satellites whose mission was to monitor changes to the global environment. Its first Earth imaging satellite, also called an Earth observation satellite, was launched in July 1972. It was originally called *ERTS-1*, which stands for Earth Resources Technology Satellite. But the name was later changed to *Landsat-1*, which is short for Land Survey Satellite. The Landsat program is considered one of NASA's greatest Earth science triumphs.

LANDSAT APPLICATIONS

Landsat data has been used by government, commercial, industrial, civilian, military, and educational organizations in the United States and throughout the world. Landsat images are routinely used by global-change researchers, farmers, forestry managers, geologists, geographers, cartographers, and oceanographers. Landsat satellites have also proven extremely useful in identifying and monitoring water supplies, assessing flood damage, planning disaster relief and flood control programs, navigating ships through icy seas, measuring air pollution, mapping forest fires, and gauging the health and likely yield of farm crops.

Over the years, the Landsat satellites have been upgraded and improved. Today *Landsat-7* continues to deliver state-of-the-art images of Earth to scientists, environmentalists, governments, businesspeople, farmers, and shippers (among other customers). Many other Earth imaging satellites have been modeled on the Landsat satellites and gather data in much the same way.

LANDSAT-1

Landsat-1 was launched on July 23, 1972, from Vandenberg Air Force Base in California. A Delta 900 multistage rocket was used to get the satellite into its circular, sun-synchronous, near-polar orbit. It orbited Earth every 103 minutes, completing 14 orbits a day. After 18 days and 251 orbits, the Landsat would cover and scan, or "swath," nearly the entire surface of Earth.

Landsat-1 had an important sensor called a multispectral scanner system, or MSS. Unlike the cameras aboard spy satellites, an MSS does not simply take a high-resolution, close-up photograph of Earth below.

As the MSS scans the sunlit surface of Earth—in 115-mile-wide (185 km) swaths of territory—it gathers Earth's reflected light through a telescope. The reflected light that is gathered reveals general details about the physical characteristics of the terrain or water below. In the completed image, the different levels of light are represented by different colors, called false colors. For instance, healthy plants show up as red; unhealthy plants show up as blue, green, or gray; and bodies of clear water are black.

On January 16, 1978, *Landsat-1* was officially taken out of service. When it was launched in 1972, it had been expected to remain operational for only a year. Instead, it remained in orbit for five and a half years and transmitted more than 300,000 amazing pictures of Earth. *Landsat-1*, the first Earth imaging satellite, proved that these spacecraft could be extremely useful in a wide range of uses: land surveying, land management, water resource planning, agricultural forecasting, forest management, sea ice movement, and cartography (mapmaking).

LANDSAT-2 AND *LANDSAT-3*

Landsat-2 was launched on January 22, 1975, so it was already in orbit when *Landsat-1* stopped working. *Landsat-2* was identical in design to *Landsat-1*. It was sent into the same type of orbit, but its orbit was timed so that

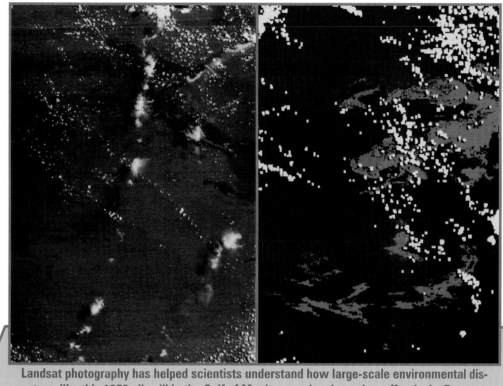

Landsat photography has helped scientists understand how large-scale environmental disasters, like this 1979 oil spill in the Gulf of Mexico, can be cleaned up effectively. Because less than one-hundredth of one percent of Earth's water is fit for human consumption, Landsat satellites continually monitor this valuable natural resource.

the two satellites could work together. Between them, they could scan the entire planet every nine days instead of every eighteen days. The two recorders on *Landsat-2* broke down, in January 1977 and May 1981, and *Landsat-2* was officially removed from service in February 1982. Like *Landsat-1*, it was designed to work for only one year, but instead continued to send hundreds of thousands of images long after its expected life span had passed.

　　Landsat-3 was launched on March 5, 1978. Identical in design to *Landsat-1* and *Landsat-2*, it also had a one-year life span, but operated far beyond that. It too was placed in a near-polar orbit and scanned the surface of the entire planet every eighteen days. The images it sent back had a resolution of 82 yards (75 meters), meaning that anything larger than 82 yards could be clearly distinguished in a photograph. Following *Landsat-3*, the management of Landsat operations was transferred from NASA to the National Oceanic and Atmospheric Administration (NOAA), a government agency in charge of describing and predicting changes in Earth's environment as well as conserving and managing America's coastal and marine resources.

LANDSAT-4 AND *LANDSAT-5*

The *Landsat-4* satellite was launched on July 16, 1982, and placed into an orbit closer to Earth than those of the previous Landsats, allowing it an even more detailed look at Earth's surface. It carried improved instruments, including a new scanner called the thematic mapper (TM). It could create images of unprecedented detail and clarity, surpassing those of the older multispectral scanner system, which was still included on board. *Landsat-4* was mainly used to monitor and manage food resources, such as crops; water resources; mineral and petroleum explorations; and land-use mapping, which is the use of aerial images to help determine the best way to develop a given

San Francisco, Oakland, Berkeley, and other cities in the Bay Area of California lie directly over an active fault line that experiences many, usually very minor earthquakes each year. Landsat photography provides valuable information about what impact this seismic activity has on the area's ecology.

area. It could distinguish soil from vegetation and deciduous trees from coniferous trees. It could also determine the moisture content of soil and vegetation and estimate when plants and crops had reached their growth peak. Like the earlier Landsats, *Landsat-4* was able to locate the presence of mineral deposits, outline areas of shallow water, and identify water that contained high levels of sediment, which are small particles of solid matter.

In August 1993, *Landsat-4* began to malfunction and had stopped gathering data due to a failure of its onboard data transmitters. As a result, *Landsat-5* was launched earlier than planned, on March 1, 1984, to insure continued data collection. Like *Landsat-4*, it too was placed in a lower orbit than previous Landsats and carried both the thematic mapper and the multispectral scanner. The thematic mapper system scanned Earth's surface for radiation. It was very helpful for scanning volcanic eruptions and the extent and health of vegetation.

Landsat-4 and *Landsat-5* lasted much longer than the two to five years that NOAA thought they would last. *Landsat-4* stopped sending images to Earth in 1993, after eleven years in service. But it remained in orbit until 2001. *Landsat-5* continues to collect and transmit data.

LANDSAT-6 AND LANDSAT-7

Landsat-5 was supposed to be taken out of service once *Landsat-6* was placed in orbit and operating correctly. *Landsat-6* carried even more advanced instruments than the five earlier Landsat systems. It was launched on October 5, 1993, but failed to reach its intended orbit and was lost forever. The plan was then for *Landsat-5* to continue operating, but only until *Landsat-7* was up and running. *Landsat-7* was launched successfully on April 15, 1999, but as of fall 2002, *Landsat-5* was still in use. NOAA planned to take it out of service in 2000, but some of the people who use Landsat images complained. They wanted to have a back-up system in place just in case *Landsat-7* ever failed.

Landsat-7 orbits at a distance of about 438 miles (705 km) above Earth. It crosses the equator at about the same time every morning, but in a different location. Like *Landsat-4* and *Landsat-5*, *Landsat-7* takes sixteen days to scan the entire planet. It

Begun in 1972, Landsat is the longest-running satellite-based Earth imaging program. This photograph of the Capitol Building in Washington, D.C., shows just how detailed Landsat photography has become.

orbits Earth, from pole to pole, 233 times during those sixteen days. Each of these orbits takes about ninety-nine minutes.

The advanced sensor on board *Landsat-7* is called the enhanced thematic mapper plus (ETM+). It is similar to the TM, but it is more advanced in many ways. For instance, the images that it transmits are of a much higher resolution, ranging in most cases from 49 to 98 feet (or 15 to 30 meters). The ETM+ can map very large areas of land, as well as monitor changes to land cover, such as deforestation, urbanization, or coastal erosion.

About 250 separate images are transmitted from *Landsat-7* to the ground station every day. When a satellite sends data (such as images) to Earth, it is called downlinking. When a ground station sends commands up to the satellite, it is called uplinking. There are four main ground stations on Earth that communicate in this way with *Landsat-7*. They are located in Sioux Falls, South Dakota; Poker Flat, Alaska; Wallops, Virginia; and Svalbard, Norway.

The Landsat ground system includes a spacecraft control center, which monitors the satellite's operations and attempts to fix any technical problems; the ground stations; and a data-handling facility and archive. Together, these facilities communicate with *Landsat-7*; control all spacecraft and instrument operations; and receive, process, archive, and distribute the data sent from the satellite.

EO-1

NASA's and NOAA's series of Landsat satellites have been providing continuous images of Earth longer than any other satellites. NASA is already hard at work on the satellite that will eventually replace *Landsat-7*. On November 21, 2000, it launched an experimental spacecraft called *Earth Observing-1*, or *EO-1*. The craft's mission was to test new Earth imaging technology and determine if it was

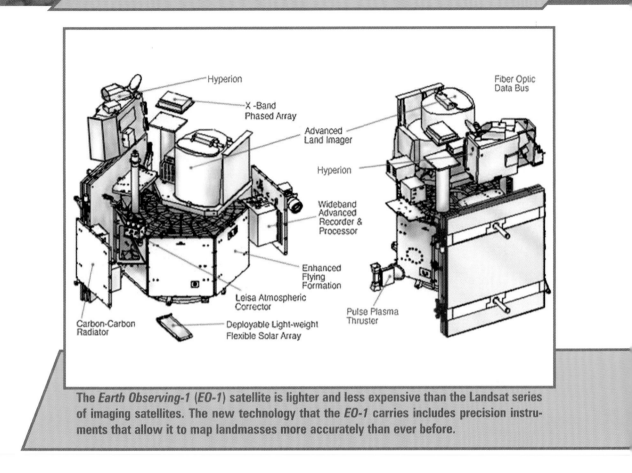

The *Earth Observing-1 (EO-1)* satellite is lighter and less expensive than the Landsat series of imaging satellites. The new technology that the *EO-1* carries includes precision instruments that allow it to map landmasses more accurately than ever before.

capable of long-term missions of the sort Landsats had been performing for almost thirty years. The hope is that the new technologies tested by *EO-1* will someday help lower the cost of future imaging missions, while providing images of greater clarity, accuracy, and detail.

On board *EO-1* are some imaging instruments that are even more advanced than the ones on *Landsat-7*. One is called the Hyperion, a hyperspectral satellite sensor. ("Hyperspectral" means it uses many more bands of the color spectrum than older multispectral scanners did,

SATELLITES TO THE RESCUE

Using environmental satellites equipped with rescue tracking equipment, the National Oceanic and Atmospheric Administration and the Russian government saved 166 lives in U.S. waters and wilderness in 2001. The NOAA satellites are part of an international search and rescue, satellite-aided tracking program referred to as Cospas-Sarsat. A cluster of satellites in both geostationary and polar orbits detect and locate emergency beacons that have been activated on ships and aircraft experiencing trouble. Engine fires, flooding, and rough, stormy seas all led to emergencies that resulted in distress calls and rescues. Of the 166 people rescued, 112 people were saved on the seas, 39 in the Alaska wilderness (often lost or stranded hikers and hunters), and 15 from downed aircraft.

—Source: NOAA News Online

vastly improving resolution and clarity.) The Hyperion system is sensitive to very minor differences in colors. When the transmissions from Hyperion are processed by computers, there are so many minor color differences that it is possible to see where one type of tree ends and another type of tree begins.

The Hyperion offers 30-meter (98-foot) resolution; an advanced land imager (ALI), a high-performance multi-spectral sensor that is also on board, produces images with a 10-meter (33-foot) resolution. *EO-1* flies right behind *Landsat-7*. It takes pictures of the same areas as *Landsat-7*, at almost exactly the same time. This way, scientists can compare the images from the two satellites and see if the

EO-1 is really doing its job and if it represents a significant improvement over the Landsat.

After a one-year testing period, ALI's image clarity was shown to be greater than that of the ETM+ on board the Landsats. The Hyperion sensor has distinguished itself as well. Where Landsat satellites had enabled researchers to identify vegetation as hardwood, softwood, and grass-lands, Hyperion provides for even more detailed and precise identification. Not only can it distinguish between red pines and red oaks, it can also identify specific types of environments, such as hardwood bogs, mixed conifers, and spruce swamplands. It can distinguish healthy grass-lands from unproductive vegetation, riverbeds from brush, and paved surfaces from dirt roads. Hyperion can even identify what type of crops are planted in a given field.

The data that EO-1 can provide and its wide range of uses have attracted great interest worldwide. Scientists from Australia's Commonwealth Scientific and Industrial Research Organization (CSIRO) have been chosen as part of an international team that will study information gath-ered by EO-1. They hope to test how useful EO-1 is for mapping and gauging the health of Australian rainforests, mangroves, deserts, and coral reefs. In addition, they hope the EO-1 images can provide important information on the quality of the nation's crops, water supply, and woodland vegetation.

CHAPTER THREE

EARTH IMAGING SATELLITES AT WORK

Earth imaging satellites are used to measure pollution, find earthquake fault lines, and monitor the surface temperature of the oceans. They are also used to watch forest fires, plan the growth of cities, and examine the mineral composition of soil and rocks. The uses for Earth imaging satellites are almost endless, and more important uses are being discovered almost every day.

NATURAL DISASTERS

Geostationary Earth imaging satellites, because they remain hovering over a fixed location on Earth, can be used to estimate rainfall in a given area during severe thunderstorms and hurricanes. This information helps predict the likelihood of flash floods. Similarly, they can estimate snowfall accumulations that help in issuing winter storm warnings and spring snow melt advisories. Geostationary Earth imaging satellites can provide weather data that allows for the early and rapid issuing of tornado warnings.

In addition to monitoring the effects of natural disasters, Earth imaging satellites can help predict where future disasters will occur. For example, by comparing this image of the Missouri River directly after a 1993 flood with a photograph of the region before the flood, scientists will be able to determine what areas along the river are likely to flood again in the future.

Earth imaging satellites can also study disaster areas and suggest remedies that will prevent similar catastrophes in the future. In the late 1970s, a prolonged drought in Sahel, a huge prairie south of the Sahara Desert in Africa, resulted in mass starvation and death. Information provided by Landsat satellites established that overgrazing by livestock had led to the spread of the desert. Areas in which livestock were fenced in and carefully tended (rather than allowed to roam free) were comparatively green and lush. Armed with this information, the inhabitants of the Sahel could try to prevent any further spreading of the desert, erosion of soil, and loss of vegetation.

ENVIRONMENTAL CHANGE

One of the important features of Earth imaging satellites is their ability to create an enormous archive of years worth of images of a single area. This allows researchers to chart exactly how a region changes from year to year. This information is particularly important to scientists studying environmental change.

One dramatic instance of environmental change recorded by Landsat satellites is of the Aral Sea, which is actually a fresh water lake, that separates the countries of Kazakhstan and Uzbekistan. Over the last thirty years, the sea had shrunk by 60 percent, but nobody knew why. Landsat images helped solve the mystery. In order to irrigate nearby cotton fields and rice paddies, farmers diverted some of the rivers that fed into the Aral, cutting off its water supply. As the water level of the lake began to drop, its chemical composition changed, becoming more salty. As a result, its fish population plummeted. As the Aral became smaller, more of the lakebed became exposed to the air. High winds began to pick up this soil and pollute the air, reducing air quality and harming crops. Without the help of the Landsat archive,

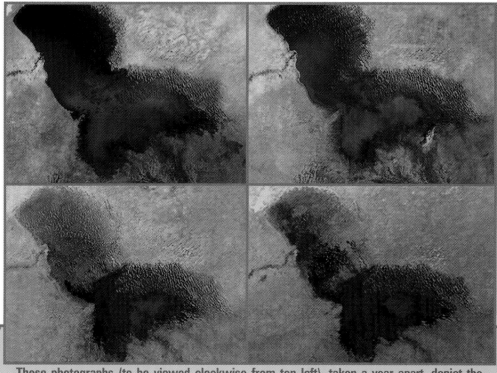

These photographs (to be viewed clockwise from top left), taken a year apart, depict the gradual drying up of Lake Chad, one of Africa's largest concentrations of drinkable water. The gradual warming of the global climate, the advancement of the Sahara Desert, and the severe water needs of the surrounding population have all helped to drain the once great lake.

this complex web of environmental change may never have been clearly understood by scientists and the residents most affected by it.

POLAR ICE

Antarctica is a continent that is in a constant state of flux, as the ice that covers it melts, refreezes, and shears off into the sea. This constant change has become a source

A VANISHING FOREST

Earth imaging satellites have provided a dramatic record of the destruction of Brazil's rainforests. Throughout the 1980s, Brazil lost 5,800 square miles (15,000 square km) of forest a year, and that pace increased in the 1990s. A rising population and the government's offer of cheap land to farmers have fueled the clear-cutting of the rainforest. Because rainforests absorb the carbon dioxide that contributes to the greenhouse effect and global warming, their health and survival are crucial to our own.

of anxiety for environmentalists who fear that global warming may lead to the melting of the polar ice caps and a catastrophic rising of sea levels. *Landsat-7* passes over Antarctica sixteen times a day. During the summer months it takes about 300 images of the continent a week. This allows scientists to spot any sudden changes to the surface that may precede a major ice shift.

While studying a Landsat image of the Pine Island Glacier in Antarctica in September of 2000 (Antarctica's winter), NASA glaciologist Bob Bindschadler noticed a 15-mile (25-kilometer) crack that did not appear in similar images taken only ten months before. This was a very quickly developing crack that seemed to indicate an imminent and major break in the Antarctic ice. It was determined that the crack had developed in only five weeks. Bindschadler predicted that a huge iceberg would shear off from the glacier (which has been shrinking since 1992) within the next eighteen months or

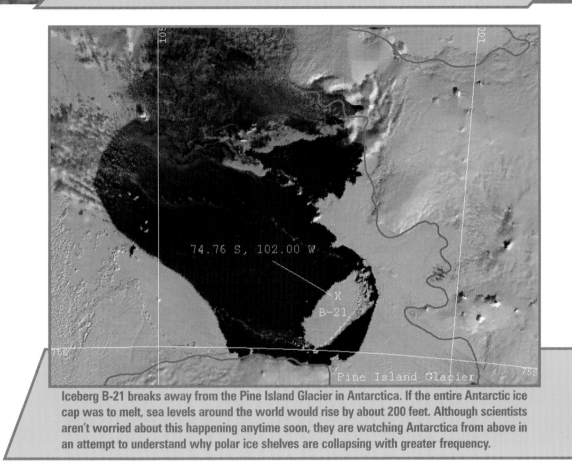

Iceberg B-21 breaks away from the Pine Island Glacier in Antarctica. If the entire Antarctic ice cap was to melt, sea levels around the world would rise by about 200 feet. Although scientists aren't worried about this happening anytime soon, they are watching Antarctica from above in an attempt to understand why polar ice shelves are collapsing with greater frequency.

so. In November 2001, his prediction came true as an iceberg 25 miles (40 kilometers) long and 9 miles (14.5 kilometers) wide "calved," or separated, from the Pine Island Glacier.

INDUSTRY

When searching for a safe location for new power plants or oil and gas pipelines, utility companies often turn to images provided by Earth imaging satellites.

Massive explosions, blackouts, and other service disruptions could result if a plant or pipeline is built along an earthquake fault line. Satellite images that identify fault and fracture zones help utility companies find the most secure locations for their plants and plot the safest, most direct routes for their pipelines. One proposed gas pipeline in Bolivia had to be redirected when Landsat data revealed that a fault line lay directly in its path.

NAVIGATION

If Earth imaging satellites had been available in 1912, the *Titanic* may never have struck an iceberg and sank, resulting in the tragic deaths of 1,517 passengers and crew. Satellite images can now clearly show ice distribution, growth, change, and movement. Shippers can use these images to determine which ports are ice free and chart courses through the clear lanes of icy waters. In 1975, a U.S. Coast Guard icebreaker, the *Burton Island*, was sailing on an Antarctic scientific expedition and became surrounded by almost solid pack ice. Using Landsat images, it managed to locate and wend its way through narrow channels of open water that were not visible to the ship's captain and crew. Without this assistance from an "eye in the sky," the ship may have become stuck until the spring thaw.

URBAN PLANNING

Images provided by earth imaging satellites have become a useful tool for both urban planners and environmentalists. Planners use them to determine the best location for new development, while environmentalists study them to chart the environmental impact of urban growth upon

In Finland's Gulf of Bothnia, an icebreaker plunges through a heavy ice field that surrounds the ship. With help from satellites, these ships have less of a chance of getting stuck in ice fields that they didn't anticipate and can't navigate.

an area. With every addition of a new road, building, or factory, the area's vegetation, water supply, animal population, and air quality are affected. Environmentalists can use this data to argue against certain kinds of development or for more ecologically sensitive construction. Planners study the satellite images to locate the most suitable spot for a given project. If they're planning a housing development or office complex, for example, planners will look for nearby water supplies, an absence of rocky soil or uneven land, and access to major traffic routes.

Earth imaging satellites do not always have such dramatic applications. Someday soon many of us may be using these images in our daily lives. City engineers studying water runoff and drainage systems may use them to determine the proportion of land covered by concrete, grass, trees, and blacktop. School and city transportation officials may study satellite images to plot the best bus routes. Farmers will be able to glance at images of their fields and quickly determine the health of their crops, helping them to decide which areas need more fertilizer or insecticide and anticipate the size and financial value of the harvest.

Before long, consulting the data produced by Earth imaging satellites may be as routine for all of us as studying a map, reading the weather report in the newspaper, or surfing the Web.

CHAPTER FOUR

THE FUTURE

Just 100 years ago, it was probably hard to imagine that one day there would be thousands of machines looking down on us from outer space. Likewise, it is difficult to predict what types of amazing inventions will be part of our world 100 years from now. Scientists who work with satellites, however, have some ideas about future technologies. Earth imaging satellites are likely to develop along two parallel tracks in the future: research satellites that study and predict things like weather, environmental change, and ocean currents; and commercial satellites that produce images to be sold to the public.

NOAA-17(M)

On June 24, 2002, a new environmental satellite—the National Oceanic and Atmospheric Administration-17(M), or *NOAA-17(M)*—was carried into a polar orbit by a Titan II rocket. It is the third in a series of five polar-orbiting operational environmental satellites (POES) that will be

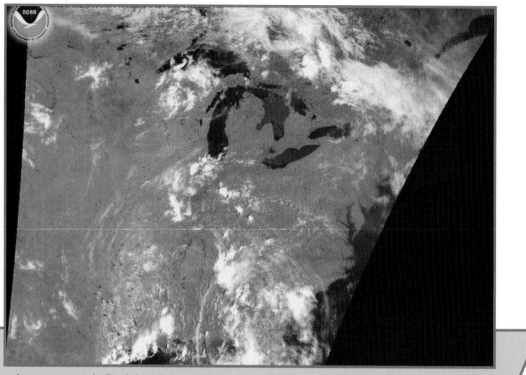

Improvements in Earth imaging satellites allow meteorologists to keep an eye on weather systems and generate more accurate short- and long-term forecasts.

put in place to offer improved weather forecasting and monitoring of environmental events around the world.

Each POES contains improved instruments that will provide better imaging and data collection than earlier generation satellites. An advanced microwave sounding unit will provide improved temperature and water vapor monitoring, especially in very cloudy conditions. An advanced high resolution radiometer will be able to distinguish between cloud cover and snow or ice on the ground, improving the images' resolution. The satellites' onboard computers feature increased memory to power these new instruments.

NOAA-17 will collect meteorological data and send it to users throughout the world to enhance weather forecasting. In the United States, the National Weather Service will use the data for its long-range weather and climate forecasts. NASA's Earth Science Enterprise (a program dedicated to understanding the total Earth system and the effects of natural and human-induced changes to the global environment) will study the information to gain a better understanding of Earth's natural processes and the extent to which they are affected or disrupted by human activity.

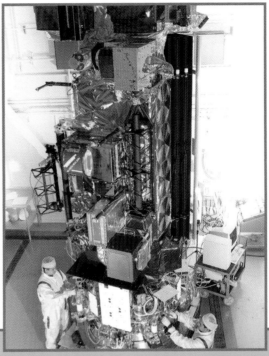

Repairing satellites once they are in orbit is costly and difficult, so the *NOAA-17(M)* is given a thorough testing by Lockheed Martin technicians before being sent into space.

AQUA

One of NASA's newest Earth imaging satellites is called *Aqua*. Its mission is to gather information about the water in Earth's system—its presence, its quality, its

Roughly $952 million has already been spent on the *Aqua* mission, designed to improve meteorological forecasts and track changes in Earth's climate. *Aqua* will also monitor how water circulates around the globe to help us gain a better understanding of water's role in climate change.

movements, its relationship to climate and climate change. The program is an international partnership between the United States, Japan, and Brazil, and the satellite is currently in a testing phase after being launched in May 2002.

The *Aqua* satellite contains six high-performance instruments, each collecting measurements on a different aspect of Earth's water system. For example, the satellite's advanced microwave scanning radiometer will produce global maps of sea surface temperatures. The onboard atmospheric infrared sounder will make highly accurate measurements of air temperature, humidity, clouds, and surface temperatures. This type of data will be extremely useful in both predicting short- and long-term weather patterns and tracing the relationship between ocean temperature and climate change. Claire Parkinson, the *Aqua* project scientist at NASA's Goddard Space Flight Center, claims in an article on NASA's Web site that, "[I]f all goes as planned, these data will lead to improved weather forecasts and a better understanding of Earth's climate system—especially the role of water in it."

AURA

Aura's mission, scheduled to begin in 2003, is to trace gasses in Earth's atmosphere. The data gathered will help scientists better understand and address issues such as global warming, the movement and distribution of pollution, and the depletion of the ozone layer. The onboard tropospheric emission spectrometer will study the troposphere, the layer of the atmosphere 10 miles (16 km) above Earth. Its main purpose will be to measure ozone levels. High ozone levels in the troposphere indicate polluted environments and are dangerous to plants, animals, and humans. Ozone levels in the stratosphere will be studied by *Aura*'s microwave limb sounder. At this atmospheric level, low levels of ozone pose a danger. Stratospheric ozone shields us from the Sun's harmful ultraviolet rays.

CLOUDSAT

The CloudSat program is composed of three satellites, due to be launched in 2003. They will be the first spacecraft to study clouds worldwide. An onboard advanced radar will create a top-to-bottom cross section of clouds. Current weather satellites can only make images of a cloud's top layers. It is hoped that by gaining a better understanding of how clouds form and are structured, we will gain insights into clouds' role in climate change.

COMMERCIAL IMAGING SATELLITES

Though the operation of spy and Earth imaging satellites was once firmly in the hands of government and military agencies, more and more photo satellites are launched by private companies, and their images are sold directly to the public. A new era of space is beginning, and images that were available only to spies and military officers with security clearance will now be available on the open market. This new access to information is both thrilling and troubling, as the information that is bought can be used for both constructive and destructive purposes.

Ikonos II, the world's first spy-quality Earth imaging satellite owned by a private corporation, Space Imaging, was launched into orbit in September 1999. *Ikonos I* failed to reach orbit the preceding April when the rocket did not separate from the satellite. Built by Lockheed Martin (a leading aerospace company) with cameras designed by Eastman Kodak, *Ikonos II* is able to transmit extremely detailed views of Earth, and these views are available to

A GROWING INDUSTRY, A SHRINKING WORLD

American companies are not the only ones getting in on commercial imaging satellites. India was the first nation to offer satellite images to the public that featured 5-meter (16.4 feet) resolution. France and Russia currently offer a similar service, and several U.S. and foreign companies are expected to launch satellites similar to *Ikonos II* in the near future. It is expected that this will soon become a $3 to $5 billion dollar industry, with most companies selling their images starting at $400 per image and up. The opportunity to make money will only increase the number of imaging satellites in space and the number of countries involved, raising troubling questions about what is being photographed, by whom, and for what purposes.

anyone who can pay for them. The resolution of its images is 1 meter (3.3 feet). This is almost as precise as military spy satellites, which are now thought to have a resolution of several inches. While *Ikonos II* will not be able to pinpoint and photograph individual people, it can identify boats, trucks, roads, and individual trees from 400 miles (644 km) up. The best resolution images are provided by its black-and-white camera, while its color camera has a resolution of 13 feet (4 meters; Kodak hopes to achieve 6-inch or 15-cm color resolution in the near future). The satellite travels at 4 miles (6.4 km) a second, allowing it to take a picture of the same spot on Earth once every few days.

Kodak sells high-resolution color and black-and-white satellite images on its Earth Imaging Products Web site. It also sells hard copy images through Space Imaging. The images cover more than 7,000 cities and towns and 600

From a distance of 423 miles (680 kilometers), *Ikonos II* snaps a photograph of the Washington Monument. Ikonos was the first privately owned satellite to sell such high-resolution photographs to the general public.

counties in the United States and Canada. Kodak expects most of its customers to be members of the architecture, engineering, construction, telecommunications, utilities, transportation, insurance, travel, and real estate industries, as well as local, state, and federal governments. A customer simply has to visit Kodak's Web site, perform a search for the appropriate image, select it, and pay for it with a credit card. The customer can choose to receive an onscreen image or a high-quality paper print. The entire process is not much different than searching for and downloading an article on a certain topic on the Internet.

CONCLUSION

As many nations begin to launch spy-quality Earth imaging satellites and consumers worldwide gain easy access to images of virtually any location on the planet, many people are beginning to express reservations about what is being photographed and into whose hands the images may fall. While the U.S. government has "shutter control" rights—the ability to shut down imaging satellites during national security crises—over American commercial satellite companies, it has no such control over foreign companies.

Many worry that terrorists will be able to buy images that give them a clear bird's-eye view of sensitive government locations, military installations, or public buildings and monuments, helping them to plan attacks. American media has been able to obtain satellite images of Chinese airbases, weapons sites in Pakistan and India, North Korea's missile installations, and the U.S. Air Force's top-secret test range. Presumably other nations and individuals have been able to obtain similar images of sensitive American territory.

The *Ikonos II* satellite captured this image of a U.S. EP-3 spy plane as it sits on a runway of Lingshui Military Airfield in China. On April 1, 2001, the EP-3 collided with a Chinese fighter jet, causing it to make an emergency landing. This satellite photo shows the level of detail of enemy territory that imaging satellites can provide. The risky missions of spy planes may not be necessary in the future if satellite imaging technology continues to improve.

Despite these concerns, those in the industry insist upon the overriding value of access to satellite images and the exciting and valuable activities they will make possible. The chief executive officer of Space Imaging, John R. Copple, believes the possible uses for commercial imaging satellites are extremely varied and outweigh any national security concerns. He imagines that pilot training will no longer occur in expensive simulators, but over the Internet, using moving images provided by Ikonos-like satellites.

Video games and educational Web sites will allow students to fly over landmarks and historical sites.

In an interview with Daniel Sorid of SPACE.com, Copple said, " [I]f you want to go anywhere in the world, you'll be able to point and click, and you'll be able to view it, find information about it. 'What are the street addresses of buildings?' If you need to know, 'How do I get from the airport to the building?' or 'Where might I stay that's close to the location I want to go?'—all that will be available through the Internet. There will be a global information database of what we call spatial data that is geo-referenced to Earth's surface, so you'll basically have an image map of the world."

Creating a complete image map of the world was the dream behind the first Earth imaging satellites. Thirty years later, we have compiled an enormous archive that essentially collects daily snapshots of Earth, allowing us to chart its changes over time. This in turn allows us to predict—and, if necessary, attempt to prevent—future changes that will affect the planet and every living thing on it. This is still the greatest promise and hope of Earth imaging satellites—the opportunity they offer to understand our world, and in understanding it, save it for future generations.

GLOSSARY

Cold War The period between the end of World War II and the 1991 dissolution of the Soviet Union, referring to the political tension and military rivalry between the United States and the Soviet Union that stopped short of actual full-scale war.

downlink Transmission of information, including images, from a satellite to Earth.

erosion The gradual wearing away of the earth's surface, caused by the action of water, wind, etc.

fuse A device, similar to the wick of a candle, that lights an explosive.

geostationary (or geosynchronous) equatorial orbit (GEO) An orbit that is 22,300 miles (35,888 km) above Earth. Satellites in this orbit travel directly above the equator at the same speed Earth spins on its axis, and therefore remain above a fixed point on Earth. To observers on Earth, it looks like geosynchronous satellites are not moving at all.

gravity The force of an object that attracts or pulls other objects toward itself.

inertia The property of a physical body remaining at rest or in motion until acted upon by an external force.

orbit Path of a satellite around its center of attraction (such as Earth's motion around the Sun, and the Moon's around Earth); to revolve in such a path.

orbital velocity The speed a satellite requires to achieve a balance between its inertia and gravity, preventing it from flying off into deep space or crashing down to Earth.

polar orbit A type of low Earth orbit that places the satellite in an orbit that circles Earth in a north-south direction—over the poles—rather than in an east-west, or equatorial, direction.

pollutant A substance that contaminates an environment.

resolution The degree of clarity of an image.

satellite An object that revolves around a larger object in a regular and predictable orbit.

uplink Transmission of information and instructions from Earth to a satellite.

FOR MORE INFORMATION

Canadian Space Agency
6767 Route de l'Aéroport
Saint-Hubert, PQ J3Y 8Y9
Canada
(450) 926-4800
Web site: http://www.space.gc.ca

Eastman Kodak Company
Earth Imaging Store
1447 St. Paul Street
Rochester, NY 14653-7112
(877) 248-4749
Web site: http://kei.kodak.com/

Federation of American Scientists
1717 K Street NW, Suite 209
Washington, DC 20036
Web site: http://www.fas.org/

Goddard Space Flight Center
Code 130, Office of Public Affairs
Greenbelt, MD 20771
(301) 286-8955
Web site: http://www.gsfc.nasa.gov

Kennedy Space Center
Public Inquiries
KSC, FL 32899
(321) 867-5000
Web site: http://www.ksc.nasa.gov/

NASA Headquarters
Information Center
Washington, DC 20546-0001
(202) 358-0000
Web site: http://www.nasa.gov/

National Oceanic and Atmospheric Administration (NOAA)
14th Street & Constitution Avenue NW, Room 6217
Washington, DC 20230
(202) 482-6090
Web site: http://www.noaa.gov/

Space Imaging
12076 Grant Street
Thornton, CO 80241
(800) 425-2997
Web site: http://www.spaceimaging.com/

The Tech Museum of Innovation
201 South Market Street
San Jose, CA 95113-2008
Web site: http://www.thetech.org

United States Geological Survey (USGS)
EROS Data Center (Earth Resources Observation Systems)
47914 252nd Street
Sioux Falls, SD 57198-0001
(605) 594-6151
Web site: http://edc.usgs.gov

United States Strategic Command
Public Affairs
901 SAC Boulevard, Suite 1A1
Offutt AFB, NE 68113-6020
(719) 554-6889
Web site: http://www.stratcom.mil

WEB SITES

Due to the changing nature of Internet links, the Rosen
Publishing Group, Inc., has developed an online list of
Web sites related to the subject of this book. This site is
updated regularly. Please use this link to access the list:

http://www.rosenlinks.com/ls/eisa/

Barrett, Norman. *The Picture World of Rockets and Satellites.* New York: Franklin Watts, 1990.

Fields, Alice. *Satellites.* New York: Franklin Watts, 1981.

Fox, Mary Virginia. *Satellites* (Inventors & Inventions). Tarrytown, NY: Marshall Cavendish Corporation, 1996.

Kallen, Stuart A. *The Race to Space.* Edina, MN: ABDO Publishing, 1996.

Mellett, Peter, and Alex Pang. *Launching a Satellite*. Portsmouth, NH: Heinemann, 1999.

Sabin, Francene. *Rockets and Satellites*. Minneapolis, MN: Troll, 1985.

Spangenburg, Ray, and Kit Moser. *Artificial Satellites*. New York: Franklin Watts, 2001.

Vogt, Gregory. *Rockets* (Exploring Space). Bloomington, MN: Bridgestone Books, 1999.

Walker, Niki. *Satellites and Space Probes*. New York: Crabtree Publishers, 1998.

Angelo, Joseph A., Jr. *The Dictionary of Space Technology*. New York: Facts on File, Inc., 1999.

"Aqua Mission Status." Earth Observing System Highlights Archive. June 24, 2002. Retrieved June 2002 (http://eospso.gsfc.nasa.gov/).

Asimov, Issac. *Exploring Outer Space: Rockets, Probes, and Satellites*. Milwaukee, WI: Gareth Stevens Publishing, 1998.

Burroughs, James. *Watching the World's Weather.* Cambridge, England: Cambridge University Press, 1991.

Commission on Physical Science Staff. *The Role of Small Satellites in NASA and NOAA Earth Observation Programs*. Washington, DC: National Academy Press, 2000.

Conway, Eric D., and the Maryland Space Grant Consortium. *An Introduction to Satellite Image Interpretation.* Baltimore. MD: Johns Hopkins University Press, 1997.

Dubno, Daniel. "Satellites Change How We See the Earth." CBSNews.com. June 3, 1999. Retrieved June

2002 (http://www.cbsnews.com/stories/1999/
06/03/tech/digitaldan/main34059.shtml).

Gurney, R.J., J.L. Foster, and C.L. Parkinson, eds.
*Atlas of Satellite Observations Related to Global
Change.* Cambridge, England: Cambridge
University Press, 1994.

Hansen, Brian. "Two Satellites Launched to Survey Earth's
Environment." Environment News Service. November
21, 2000. Retrieved June 2002 (http://ens-news.com/
ens/nov2000/2000-11-21-15.asp).

Hill, Janice. *Weather from Above: America's
Meteorological Satellites.* Washington, DC:
Smithsonian Institution Press, 1991.

Kidder, Stanley Q., and Thomas H. Vonder Haar.
Satellite Meteorology: An Introduction. San Diego,
CA: Academic Press, 1995.

"Landsat Data." United States Geological Survey.
December 1997. Retrieved June 2002
(http://mac.usgs.gov/mac/isb/pubs/factsheets/
fs08497.html).

"The Landsat Program." NASA. April 22, 1999.
Retrieved June 2002 (http://www.earth.nasa.gov/
history/landsat/).

"The Landsat Satellites: Unique National Assets."
NASA Facts Online. March 1999. Retrieved June
2002 (http://pao.gsfc.nasa.gov/gsfc/service/
gallery/fact_sheets/earthsci/landsat/landsat7.htm).

Luther, Arch. *Satellite Technology: An Introduction*.
Burlington, MA: Focal Press, 1997.

"NASA's Earth Observing Technology Satellite Proves a Success." Earth Observing System Highlights Archive. June 24, 2002. Retrieved June 2002 (http://eospso.gsfc.nasa.gov/).

"NOAA-17 (M) Environmental Satellite Successfully Launched." Earth Observing System Highlights Archive. June 24, 2002. Retrieved June 2002 (http://eospso.gsfc.nasa.gov/).

"NOAA Satellites Help Rescue 166 People in U.S. in 2001." NOAA News Online. February 1, 2002. Retrieved June 2002 (http://www.noaanews.noaa.gov/).

Parkinson, Claire L. *Earth from Above: Using Color-Coded Satellite Images to Examine the Global Environment*. Sausalito, CA: University Science Books, 1997.

"This Planet Earth: The Vision and Majesty of NASA's Remote Sensing Legacy." NASA Goddard Space Flight Center. April 19, 2001. Retrieved June 2002 (http://www.gsfc.nasa.gov/gsfc/earth/imaging/landsat.htm).

Sorid, Daniel. "Ikonos Safely in Orbit." SPACE.com. Sept. 24, 1999. Retrieved June 2002 (http://www.space.com/missionlaunches/launches/ikonos_launch_990924.html).

Sorid, Daniel. "A View of the Future of Viewing Earth." SPACE.com. Sept. 16, 1999. Retrieved June 2002 (http://www.space.com/peopleinterviews/spaceimaging_ceo.html).

Stott, Carole. *Space Exploration*. New York: Alfred A. Knopf, 1997.

Venere, Emil. "Future Imaging Satellites to Have Everyday Applications." Purdue News. October 1999. Retrieved June 2002 (http://www.purdue.edu/UNS/html4ever/990723.Landgrebe.spy.html).

INDEX

CREDITS

ABOUT THE AUTHOR

Jan Goldberg is an experienced educator and the author of more than thirty-five nonfiction books and hundreds of educational articles, textbooks, and other projects.

PHOTO CREDITS

Cover, pp. 5, 35, 48, 50 © AFP/Corbis; pp. 8, 10, 11 © Bettmann/Corbis; pp. 13, 44 © AP/Wide World Photos; pp. 16, 19, 29, 42, 43 © Goddard Space Flight Center/ NASA; pp. 23, 25, 33 © Corbis; p. 27 © Reuters NewMedia, Inc./Corbis; p. 37 © Reuters/Timepix; p. 39 © Nik Wheeler/Corbis.

DESIGNER

Thomas Forget